# DRY CLIMATES

## Keith Lye

RSVP®

RAINTREE
STECK-VAUGHN
P U B L I S H E R S
The Steck-Vaughn Company

*Austin, Texas*

# THE WORLD'S CLIMATES

# COLD CLIMATES
# DRY CLIMATES
# EQUATORIAL CLIMATES
# TEMPERATE CLIMATES

Published by Raintree Steck-Vaughn Publishers, an imprint of Steck-Vaughn Company

**Library of Congress Cataloging-in-Publication Data**
Lye, Keith.
Dry climates / Keith Lye.
    p.    cm.—(The world's climates)
    Includes bibliographical references and index.
    Summary: Examines the many different types of dry climates, from deserts to semiarid regions.
    ISBN 0-8172-4828-5
    1. Arid regions climate—Juvenile literature.
    [1. Arid regions.  2. Deserts.]
    I. Title.  II. Series: Lye, Keith. World's climates.
    QC993.7.L94   1997
    551.6915'4—dc20          96-31161

Printed in Italy. Bound in the United States.
1 2 3 4 5 6 7 8 9 0 01 00 99 98 97

Cover picture: Monument Valley, Robert Harding Picture Library

**Picture acknowledgments**
Britstock-IFA p14; Carol Kane pp 5, 15, 18, 22, 43; Corbis-Bettman p 40; FLPA pp 16–17, 32, 33, 41; Hutchinson Library p 29; Panos Pictures p 27; Robert Harding Picture Library pp 5, 6, 7, 9, 11, 15, 17, 19, 20, 23, 24, 25, 26, 29, 30, 32, 34, 36, 38, 39, 41, 42, 43; Tony Stone Images pp 1, 16; Trip pp 13, 19, 21, 30, 31, 35, 36–37, 37.

All illustrations by Timothy Lole except p39, Alistair Wilson

# Contents

# Arid and Semiarid Regions

## CLIMATIC REGIONS

Climate is the usual, or average, weather of a place. It determines what plants and animals are found in an area, and it influences how people live. The two main factors that determine climate are temperature and precipitation. Precipitation includes rain, snow, sleet, hail, dew, and frost — in fact, all forms of moisture that come from the air.

Temperature is a major factor in determining cold climates, hot equatorial climates, and temperate climates, but one group of climate regions is based on precipitation, or rather the lack of it. These are the dry climates, which occur in hot, temperate, and cold regions.

Hot and cold deserts, which have an arid (extremely dry) climate, cover about 12 percent of the earth's land surface. Places with semiarid climates cover a further

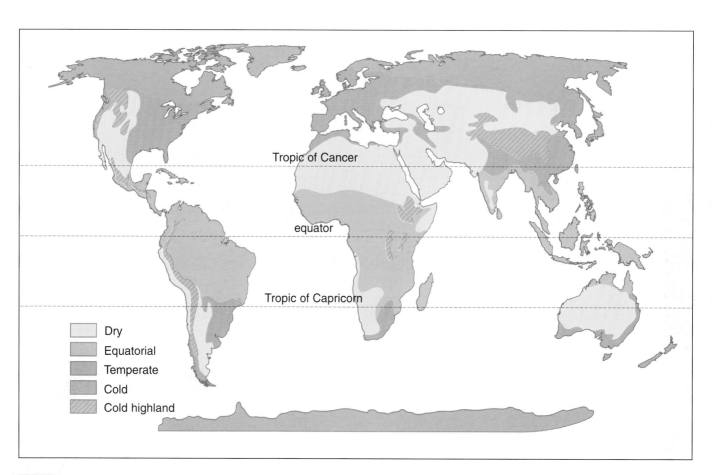

Tropic of Cancer

equator

Tropic of Capricorn

Dry
Equatorial
Temperate
Cold
Cold highland

14 percent of the earth's land surface. In semiarid regions, the average yearly precipitation is greater than in deserts, but it is still so low that forests cannot grow. Grasses are the most common plants in semiarid regions, which include the prairies of North America, the steppes of Europe and Asia, the pampas of South America, and the downlands of Australia and New Zealand.

The lands around the North and South poles also have little precipitation and can be described as dry climate regions. However, the major climatic features in polar regions are the extremely low temperatures, so experts usually classify these regions as cold rather than dry climates.

Regions with desert climates are among the most hostile on earth. Few plants grow there, and animals must be able to go without water for long periods. Few people live in deserts except around oases, such as wells or springs.

Semiarid regions support a wider range of plants and animals than deserts do. They are thinly populated, but they are used for ranching or growing grains, such as wheat.

**Far left** The main climate regions of the world

**Left** Deserts are barren and hostile environments.

**Below** The prairies of North America are used for wheat farming.

## WHAT IS A DRY CLIMATE?

Places with arid (desert) climates have a total yearly precipitation of less than 10 inches (250 mm) a year, whereas places with semiarid climates have from 10 inches to 36 inches (250–900 mm). Another important factor in dry areas is the balance between the amount of rain that falls and the amount that is lost by natural processes.

Water is lost in two main ways. Some of the rain that falls in hot regions is quickly evaporated on the ground by the sun's heat. Evaporation is the change that occurs when water (a liquid) is changed by heat into water vapor (a gas). The air over hot deserts

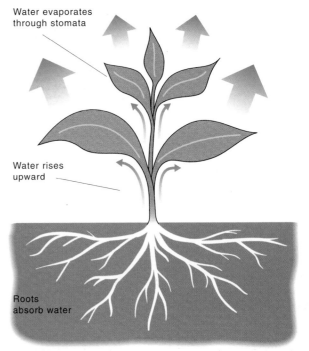

Water evaporates through stomata

Water rises upward

Roots absorb water

often contains huge amounts of water vapor, but clouds are rare; clouds form only when warm air is chilled. Cooling of the air makes the invisible water vapor change back into the billions of liquid water droplets that form clouds.

6

**Left** *The skies above the Western Desert in Egypt are cloudless.*

**Right** *The thick stems of cacti store water. These are on Gran Canaria in the Canary Islands.*

**Below left** *How water passes through the roots to the leaves of a plant*

Water is also lost by plants through a process called transpiration. This occurs because plants give off water vapor through the tiny stomata (pores) in their leaves. Transpiration is a vital part of plant growth because it keeps water moving from the roots, which get water from the soil, through the stems and up to the leaves. Plants sometimes lose water quickly through transpiration. For example, a single corn plant can lose a gallon of water through its leaves during the course of one hot day. To survive in deserts, plants must have ways of reducing transpiration and conserving water; otherwise, they will die.

Places with semiarid climates may be dry and barren if the evaporation and transpiration rates are high. Other places with less precipitation, but with low evaporation and transpiration rates, may be much less barren.

# HIGH AIR PRESSURE

The atmosphere around the earth is always on the move. Movements of air, including the winds that blow across the earth's surface, have a great influence on climate.

Winds may blow air sideways, but currents of air also move up and down. Around the equator, the sun heats the land. (The equator is the imaginary line that runs around the earth, halfway between the two poles, dividing the world into the Northern and Southern Hemispheres.) In turn, the land heats the air above the ground. The hot air rises in fast currents, creating a zone of low air pressure called the doldrums. The rising air contains invisible water vapor and cools as it rises. Because cool air cannot hold as much water vapor as warm air, the vapor turns into tiny water droplets, forming thunderclouds, which produce rain.

The rising air finally spreads out north and south in the upper atmosphere. It sinks back to the surface around latitudes 30° north and 30° south. The sinking air presses down on the surface, creating zones of high air pressure called anticyclones. Rain is rare and desert climates are common in zones where the air is sinking—the opposite of what happens when hot air rises.

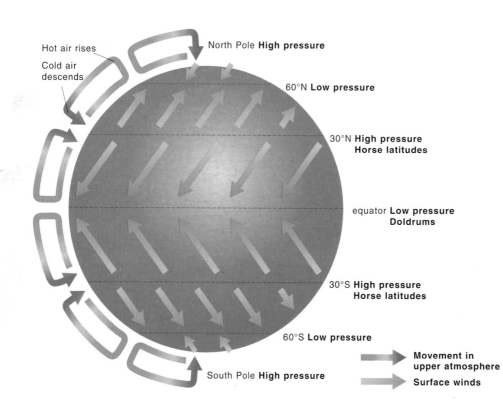

Hot air rises
Cold air descends

North Pole **High pressure**

60°N **Low pressure**

30°N **High pressure**
**Horse latitudes**

equator **Low pressure**
**Doldrums**

30°S **High pressure**
**Horse latitudes**

60°S **Low pressure**

South Pole **High pressure**

➡ **Movement in upper atmosphere**
➡ **Surface winds**

**Left** This diagram shows how air circulates in the atmosphere.

**Right** Alice Springs, a town close to the Tropic of Capricorn (23$\frac{1}{2}$° south), in central Australia

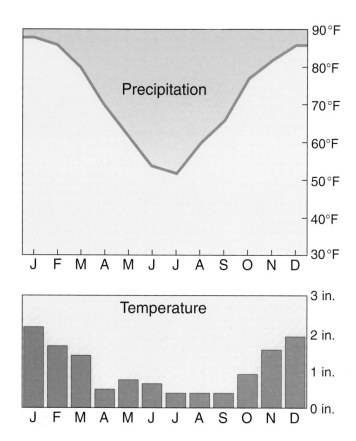

These temperature and rainfall graphs show the dry climate of Alice Springs, Australia.

High air pressure zones from about 20° to 30° north and south are called the horse latitudes. These zones were probably named in the days of sailing ships, when horses were taken from Europe to the Americas. Ships were often becalmed because of the lack of winds in the middle of the horse latitudes. When this happened, horses had to be thrown overboard so the ships could move faster.

Winds blow north and south from the horse latitudes. Trade winds carry air back across the land toward the low pressure doldrums. Other winds carry air toward the poles.

# LAND AND SEA

The world's largest deserts are in regions where air is sinking. In the Northern Hemisphere, they include the dry regions of northern Mexico and the southwestern United States, the Sahara in northern Africa, and the deserts of southwestern Asia.

In the Southern Hemisphere, they occur in central Australia, southern Africa, and southern South America.

Other forces also help create arid and semiarid climates. Many mountain ranges lie in the paths of moist, warm winds that blow from the sea. The winds are forced to rise and the air cools, causing

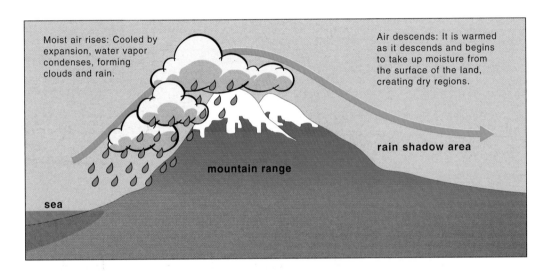

Moist air rises: Cooled by expansion, water vapor condenses, forming clouds and rain.

Air descends: It is warmed as it descends and begins to take up moisture from the surface of the land, creating dry regions.

rain shadow area

mountain range

sea

**Left** This diagram shows a rain shadow area.

**Right** Coastal desert in Algeria

**Below** A map of warm and cool ocean currents in the southwestern Pacific and south Atlantic

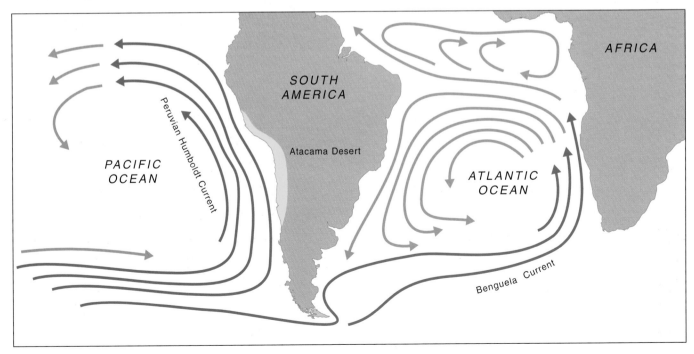

AFRICA

SOUTH AMERICA

Peruvian Humboldt Current

Atacama Desert

PACIFIC OCEAN

ATLANTIC OCEAN

Benguela Current

invisible water vapor to condense (change from a gas into a liquid) and forming clouds made up of tiny water droplets. Rain and snow fall from these clouds on the mountain slopes.

Beyond the crest of the mountains, the winds blow down the other side of the range. The air gradually gets warmer and it evaporates (takes up moisture from the surface). These drying winds create dry climate regions called rain shadow areas. The prairies of North America and the dry grasslands of Patagonia in South America both lie in rain shadow areas.

Ocean currents also affect climates. Warm surface currents flow from the tropics toward the poles, while cold currents flow from polar regions toward the tropics. Cold currents chill winds that blow across them. Water vapor in the cooling air is condensed into sea fog made up of masses of water droplets. After the winds pass over the shore, they are warmed by the land and become drying winds. Cold onshore currents are responsible for two extremely dry regions: the narrow Peruvian and Atacama Desert of west-central South America, and the Namib Desert in southwestern Africa.

# Desert Climates

## HOT DESERTS

The hot deserts, which lie roughly from 20° to 30° north and south of the equator, contain areas where daytime temperatures soar. The highest ever recorded temperature in the shade was 136.4°F (58°C). It was measured in 1922 at Al-Aziziyah, south of the Libyan capital, Tripoli. The highest recorded temperature in the United States (134.6°F or 57°C) was recorded at Death Valley, in California.

Nights in these areas of intense daytime heat are often cold. In the Algerian town of In Salah, in the middle of the Sahara, daily temperatures can vary by 100°F (55°C), (from 127°F [52°C] by day to 27°F [−3°C] at night). Desert travelers need clothes that will keep them warm at night.

Some deserts have hot summers and cold winters. For example, the deserts of Afghanistan and Iran have hot, dry summers, but snow falls in winter. Frosts also occur. An even colder region is the Gobi Desert in southern Mongolia and China. In places, winter temperatures sometimes fall below −4°F (−20°C); in summer temperatures soar to 85°F (30°C) or more.

Hot deserts contain extremely dry areas. The world's driest place is in the Atacama Desert, along the Pacific coast of Chile. Here the average yearly rainfall is only about 0.004 inch (0.1mm).

The warm air over hot deserts feels dry and clouds are rare. Yuma, Arizona, in the southwestern United States, is one of the

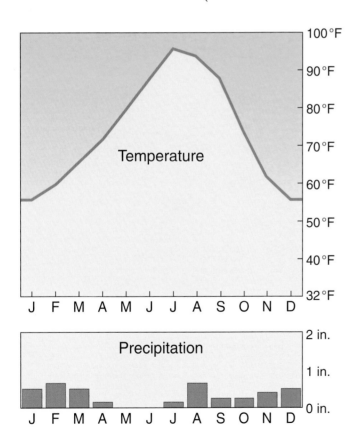

*Climate graphs for Yuma, Arizona, show little rain but high temperatures.*

**Above** This desert near Yuma, Arizona, is one of the world's sunniest places with more than 4,000 hours of sunshine annually.

**Left** Rainfall at selected desert towns located in different continents

world's sunniest places. Its average annual sunshine is about 90 percent of daylight hours.

Strong winds are another feature of hot desert climates. They lift dust and sand above the surface, creating dust storms and sandstorms. These storms are feared by desert travelers, who can easily lose their way if caught in one.

# CLIMATE AND SCENERY

The climate of deserts helps to shape the scenery. For example, the outer layer of a rock expands as it is heated during the day, but at night it contracts. These daily changes crack the outer layers of the rocks, which slowly peel away like the outer layer of an onion.

Strange as it may seem, water also molds desert scenery. The rainfall in deserts is unreliable, and much of it occurs during violent storms that affect fairly small areas. At Helwan in the Sahara, a town south of Egypt's capital, Cairo, a total of 31 inches (780 mm) of rain was recorded over a 20-year period. That means that Helwan had an average yearly rainfall of only 1.55 inches (39 mm), yet seven storms accounted for nearly 8 inches (200 mm) of the rainfall over this period. During one of the storms, more than 1.6 inches (40 mm) fell, causing flooding that ruined crops and damaged buildings.

Sudden storms in deserts create torrents of water that sweep along normally dry valleys called wadis. Sometimes people sleep in wadis, where they are protected from the wind. But faraway storms may cause walls of water to advance down the wadis, drowning the people who have no time to escape. The floods sweep away sand and stones, enlarging the wadis. Although such storms are rare, they have a great effect on shaping the scenery.

For most of the time, deserts are dry, and during these periods the

winds have a great effect. Winds are important because they blow loose dust and sand into the air. Windblown sand smooths and wears away rock, much like sandblasters that are used to clean the outsides of dirty city buildings. Windblown sand undercuts cliffs, hollowing out caves. It also carves rocks into strange shapes.

**Above**  These rocks in Nigeria have been eroded by the wind.

**Left**  A wadi in the Jordanian desert

**Right**  Rainbow Bridge, Utah, has been created by a combination of wind and water erosion.

# DESERT PLANTS

Water is often present in baking hot deserts, although it is hidden away inside the rocks. Plants that live in the desert have to be able to use this water to survive.

Some plants, such as the North American mesquite bush, are drought-tolerant. Their roots can tap water 160 feet (50 m) or more below the surface.

Other plants, such as cacti, have

**Above** Flowering mulla mulla plants in the Uluru National Park, Australia

**Above right** The Welwitschia plant has adapted to desert conditions and can live up to 2,000 years.

**Left** The Mesquite Flat Dunes in Death Valley, California, were named after their mesquite plants.

shallow roots that spread far and wide, collecting moisture from a large area. These plants, which are called drought resisters, store water in their trunks and branches. The saguaro (giant cactus) can store 6 to 8 tons of water.

Another group of drought-resisting plants, the euphorbias, or spurges, contain a sticky milky substance called latex in their stems. This substance is highly resistant to evaporation.

Some plants conserve water by closing their stomata during the day; others have waxy surfaces to reduce water loss.

Some plants are called drought evaders. Their seeds can lie around in deserts for years, but when a sudden storm occurs, they grow rapidly and scatter their seeds within 6 to 8 weeks of sprouting. This happens occasionally in dry parts of Australia, where only a few grasses and hardy shrubs normally grow. After a heavy downpour, these barren areas suddenly become colorful seas of flowering plants. Then, the plants die and the land reverts to desert.

One strange plant, the welwitschia, lives in the Namib Desert. It has two long, leathery leaves that are often shredded by windblown sand. It gets water from the tiny droplets that condense on its leaves when sea fog blows over the land.

# DESERT ANIMALS

Animals have developed many ways to survive in deserts. To avoid overheating, many insects and reptiles, such as snakes and lizards, hide during the day and are active only at night or early in the morning and evening. They spend the day in burrows, where they are protected from the burning sun. A black beetle in the Namib Desert drinks in an unusual way. It stands on its head, facing the wind. Water droplets condensed from sea fog form on its body and drip down to its mouth.

Only a few large animals can live in hot deserts. That is because water evaporates from their bodies to keep them cool and has to be constantly replaced. Some animals, such as red kangaroos in central Australia, keep cool by licking their bodies to encourage evaporation. They get water by eating moist plants, which they travel long distances to find.

A rare antelope, the addax of North Africa, seeks shade in caves or under overhanging rocks during the day. It gets moisture from its diet of grasses.

Camels can be used to carry heavy loads and can go for long periods without drinking. People once thought that the hump of a camel contained water, but this is incorrect. The hump is made of fat, and it provides a food reserve when food is in short supply.

Desert birds visit oases to drink. The sandgrouse of Africa and Asia nests in dry desert areas. When it visits a source of water, it soaks its breast feathers in the water. When it returns to its nest, its wet feathers help to keep its eggs from overheating. After the young have hatched, they get water by sucking it from their mother's feathers.

**Above left** A camel in its desert environment

**Above right** Red kangaroos live in both the central Australian deserts and the surrounding grasslands.

**Right** The large ears of the bat-eared fox help to keep the animal cool.

# Climate and Desert People

## HUNTERS AND GATHERERS

Not many people live in deserts because the climate makes life difficult. The only places with large numbers of people are oases. An oasis has a spring, where water flows on to the surface, or a well that has been dug down to water reserves in the rocks below. The largest oasis is the Nile Valley in Egypt, where little rain falls.

**Right** This Aboriginal man in Australia has painted his face in a traditional way.

**Left** A Bushman in the Kalahari Desert hunting with a bow and arrow

**Below** This map shows where Bushmen live.

The Nile gets water from rivers that rise in the wet highlands of East Africa. The Nile has most of Egypt's population.

In the past, some groups of people hunted animals and gathered plant foods in deserts.

The Bushmen of southern Africa live mainly in the Kalahari Desert in Botswana and Namibia. They live in small groups of about 25 people. Each group moves around in search of foods such as berries, nuts, roots, and seeds; hunters kill animals with poison-tipped arrows.

The Bushmen have great knowledge of the Kalahari. They know where to find roots and tubers that contain water, and places where they can suck up water through straws from beneath the surface. The Bushmen live in harmony with nature, but large numbers have now had to abandon their old way of life. The desert has been fenced in by cattle farmers, making hunting very difficult.

Some Australian Aboriginal people also lived in the deserts of the interior. They got their food by hunting and gathering. With their deep knowledge of the desert and its plants and animals, they lived without harming the land. These desert nomads (wanderers) were deeply religious and artistic, as shown by their superb cave paintings. However, like the Bushmen, many Aboriginal people have abandoned their traditional way of life and adopted European ways.

# DESERT NOMADS

Some desert people own animals, such as camels, cattle, goats, and sheep. They live a nomadic life, moving from place to place searching for pasture for their animals. These people are called pastoral nomads.

The main group of pastoral nomads in the central Sahara are the Tuareg, whose name means "people of the veil." Among the Tuareg it is the men, not the women, who are veiled. They cover their faces by winding their turbans around their heads so that only their eyes are visible. Their turbans and their long, often blue, robes give them protection against the sun and windblown sand. The women wear head scarves, but their faces are not covered.

The Tuareg are divided into several groups. The Northern Tuareg, who live in the Sahara, find pasture in the mountains that rise above the desert plains. The Tuareg once led camel caravans across the Sahara, trading with the people south of the Sahara.

Most Tuareg now live in the Sahel region, south of the Sahara. Many of them have given up their wandering life and live in permanent settlements.

**Left** Tuareg tents are usually made of goat skin or wool.

**Top right** Bedouin people use camels for transportation.

**Bottom right** Baluchi nomads with their animals, in Belaw Pass, Pakistan

That is also true of the Bedouin people who live in the Arabian peninsula. These people were famous for their hospitality to any traveler they met in the desert. Today many Bedouin have moved into permanent homes where they can obtain modern benefits, such as education and health services.

The Baluchi people of southeastern Iran and southwestern Pakistan grow crops in areas where there is enough water for irrigation, but the main occupation is sheep farming. Some Baluchi remain pastoral nomads, who are always on the move, undertaking dangerous journeys across mountains and deserts.

# LIFE IN COLD DESERTS

Some deserts lie in the interior of Asia, far from any seacoast. They include the Karakum in Turkmenistan, the Kyzylkum in Uzbekistan, the Taklimakan in western China, and the Gobi Desert, which lies in southern Mongolia and parts of northeastern China. These deserts lie far to the north of the hot desert zone, which stretches through the Middle East as far as the Thar Desert on the Pakistan-India border.

The deserts in the interior of Asia are arid, mainly because they are so remote from the moist winds that blow from the sea. Because of their northerly position, they suffer extreme temperatures, with hot

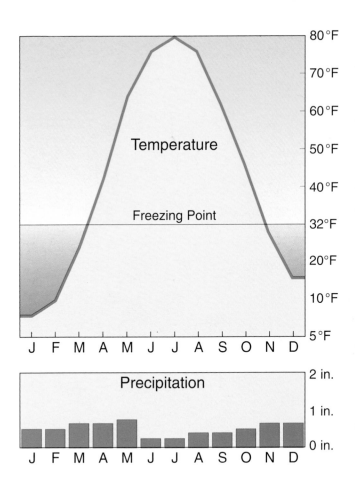

*Extreme winter and summer temperatures can be seen in these climate graphs for Novokazalinsk, in Kazakhstan.*

**Left** *Mongolians inside their yurt (a wooden framed tent covered by thick felt)*

**Above right** *A small oasis village in the Karakum Desert, Uzbekistan*

summers and bitterly cold winters. The Gobi is the most northerly of these inland deserts. Here, temperatures sometimes rise above 100°F (45°C) in summer and fall to −40°F (-40°C) in winter. Because the climate is so unpleasant, few people live there.

Mongolia, a landlocked country, which contains a large part of the Gobi, is a dry place. The mountains are the wettest areas, with an average yearly precipitation of from 15 to 20 inches (375–500 mm), some of it falling as snow during the long winter. On the lowlands, the average annual rainfall is generally 5 inches (120 mm) or less.

Many Mongolians are pastoralists who raise large herds of sheep and other animals. They once lived as nomads, using large, warm tents called gers or yurts, made of felt. Warm clothing is essential in the winter, when strong cold winds sweep across the bitterly cold, treeless plains. Most Mongolians now live in cities or on large ranches with small towns in the center. Milk, cheese, and butter are their main foods.

# LIFE ON THE EDGE OF DESERTS

Climatic regions do not have exact boundaries. Around the world's deserts, the bare land gradually merges into dry grasslands.

One well-known semiarid region called the Sahel lies to the south of the Sahara in North Africa. It extends across Africa, from southern Mauritania and Senegal on the Atlantic Ocean to Sudan on

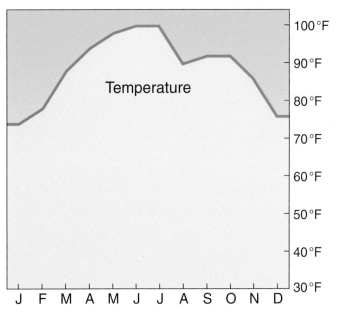

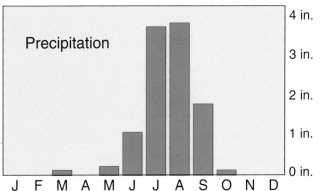

*These climate graphs show the semiarid climate of Timbuktu, Mali.*

the Red Sea. The Sahel has an average yearly rainfall ranging from about 6 inches to 16 inches (150–400 mm).

The rainfall is usually enough to support grasses, shrubs, and scattered trees. The drier areas are grazing land for nomadic pastoralists, including Tuareg, while farmers grow crops in the wetter parts of the south.

In the early 1960s, when rainfall was abundant, the people of the Sahel prospered. Their herds of animals grew larger, and the farmers increased crop production. However, the rainfall is unreliable and severe droughts hit the area starting in the late 1960s. Vast areas of grassland were laid bare by the large herds of animals, and the continuing droughts year after year turned dry grassland into barren

desert. The Sahara began to expand southward, in a process that is called "desertification."

Desertification is caused partly by natural climatic changes, but human misuse of the land is also important. People killed off grasses by grazing overlarge herds of animals, cutting down trees and shrubs for firewood, and plowing up the land. In the 1970s and 1980s, desertification led to the deaths of many animals. Their owners starved and millions of people died. In the 1980s, experts worked with the people of the Sahel to restore the region by planting trees and grasses to hold the soil in place. The droughts ended, and by the early 1990s, the Sahara began to retreat.

# Semiarid Climates

## DRY GRASSLANDS

Dry grasslands can be found in the middle latitudes—that is, the region that lies between the low latitudes of the hot tropics and the high latitudes of the polar regions.

The middle latitudes also have what are called temperate regions; such areas are not as dry as these grasslands are.

Dry grasslands have an average annual rainfall of 10 inches to 36 inches (250–900 mm). The main reason they are dry is that they lie mostly in the hearts of continents, far from the effect of moist winds that blow from the sea.

Although they have a fairly short growing season and the rainfall is often unreliable, these grassland regions have been greatly changed by human activity and have become important food-producing regions.

The drier grasslands are called steppes. They include the vast region that stretches from the Ukraine and southwestern Russia into central Asia and much of the Great Plains of North America, between New Mexico and Alberta, Canada. Here,

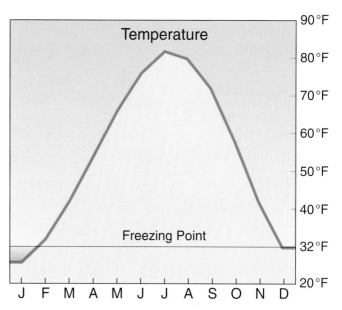

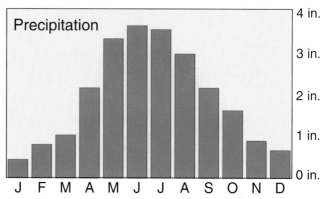

*Temperature and rainfall graphs for Dodge City, Kansas, show the semiarid climate of a prairie region.*

the average annual rainfall is less than 20 inches (500 mm). The term *steppe* is also sometimes used for dry grasslands in the tropics, and these dry grasslands are often used for grazing animals.

*Left* Sheep farming is common on the pastureland south of Adelaide, Australia.

*Below* The high veld of Drakensberg, South Africa, is a dry grassland.

The prairie is another type of dry grassland. The average annual precipitation on the prairies (from about 20 inches to 36 inches (500–900 mm) is higher than on the steppes. The soils of the prairies are fertile, so they can be used for growing wheat and other grains.

One prairie region runs through the North American Middle West, east of the Great Plains. This region extends from southern Saskatchewan in Canada to central Texas.

Other dry grasslands include the pampas of Argentina, the veld of South Africa, and the downlands of southeastern Australia. Another grassland region is the Canterbury Plains on South Island, New Zealand.

# PLANTS

Grasses are among the world's most successful plants. They grow everywhere between the tundra and the tropics. In all, they cover nearly a quarter of the earth's land surface. Their success is partly due to their lightweight seeds, which are easily carried by the wind and can spread over a wide area. Grasses are shallow-rooted plants that need frequent showers during the growing season, but the tangled grass roots hold moisture even when the surface is dry. Grasses survive fire because fire destroys only the top part of the plant, not the larger part that lies below the surface. Long droughts in the growing

season harm most grasses, but some species, such as Mitchell grass in the steppe regions of southern Australia, are drought-resistant. From 8 to 12 inches (20–32 cm) in height, the short Mitchell grass is useful as fodder for animals.

steppe has a continuous cover of grasses reaching heights of 6 to 12 inches (15–30 cm). There are also many flowering plants, such as crocuses, hyacinths, and irises.

Prairies typically have a thick cover of grasses with many species, including the 6-foot-high (1.8 m) Big bluestem, Indian grass, and wild rye, which grow in wetter regions. However, grasses still reach heights of 24 inches (60 cm), even in the eastern part of the North American prairie, where the land merges into the drier steppe of the Great Plains region. The prairies also contain many flowering plants, such as asters, sunflowers, and the prairie rose. Ribbons of trees grow along the rivers that wind across the prairie.

The types of grasses that grow depend on the rainfall. In the drier steppe regions, the grasses are lower than on the wetter prairies. In the driest steppe areas, the grasses grow in clumps, with bare ground in between. However, most

**Above** The dry climate of the North American Prairie is ideal for wheat farming. These wheat fields are in Alberta, Canada.

**Left** Grasses are farmed all over the Canadian Prairie.

**Right** Sheep are farmed in Canterbury, New Zealand.

# ANIMALS

The grasslands once supported large numbers of grazing animals. The steppes of Europe and Asia had saiga antelopes and tarpans (wild horses). North America had vast herds of bison (also called buffalo) and pronghorns, while the South American pampas had guanacos and large, flightless birds called rheas. The settlement and farming of the dry grasslands have greatly reduced the numbers of these animals, and some have been hunted almost to extinction.

The grasslands contain a wide variety of small mammals, such as jackrabbits (a kind of hare), marmots, rabbits, ground squirrels, susliks, and voles. Many of them use speed to escape predators such as coyotes, foxes, and skunks. Others, such as mice and prairie dogs, hide in underground burrows. Some birds, including buzzards, hawks, and owls, hunt small mammals; others eat insects such as

**Right** *The pronghorn of North America, once close to extinction, has recovered dramatically.*

**Right** *A koala from Australia*

**Far right** *A guanaco grazes in the mountains of Patagonia, Chile.*

grasshoppers and leafhoppers. Prairie birds include grouse, meadowlarks, quail, and sparrows, while great bustards, demoiselle cranes, and plovers are familiar birds on the steppes. Snakes include the rattlesnake in North America, with vipers and whipsnakes being common on the Eurasian steppes.

Animals of the dry grasslands must have ways of coping with the climate. The only defense that saiga antelopes have against the cold winters is a layer of fat under the skin. They huddle together in sheltered areas.

Australian marsupials can survive in areas where domestic animals, such as sheep, would starve. They can also cope with the lack of water, and many get moisture from their food. The koala's name comes from an Aboriginal word meaning "the animal that does not drink." In fact, koalas do drink occasionally, but most of the moisture they need comes from the juicy eucalyptus leaves they eat.

# THE NORTH AMERICAN PRAIRIE

When Europeans first arrived in North America in the sixteenth century, the prairies and the Great Plains were occupied by Native Americans, called the Plains Indians. They included tribes such as Blackfoot, Cheyenne, Crow, Pawnee, and Sioux. Many of these people lived a nomadic life, hunting the bison. At first, they pursued the animals on foot. Then, Spanish settlers introduced horses to North America in the seventeenth century, and the Plains Indians learned how to ride and hunt on horseback. The bison supplied most of the people's needs through their meat, their hides (which were used to make clothes and tents), their sinews

(used as thread), their bones and horns (used to make tools and utensils), and their manure (which was burned as fuel).

From the 1840s, European pioneers started to push through Native American territory in search of gold and new lands to the west. Wars eventually broke out as the Plains Indians struggled to keep their hunting lands, but the United States Army defeated them in the 1860s and 1870s. Many Plains Indians died in the wars, and the survivors were confined to special areas called reservations. The Plains Indians' way of life was finally made impossible when the vast herds of bison were slaughtered by hunters paid by the United States government. In 1850, the herds of bison numbered about 20 million; by 1889, only 551 remained in the United States.

In a short time, the prairies and Great Plains were developed as cattle ranches and huge grain farms. But overgrazing and plowing harmed many areas, which became infertile. When droughts occurred, strong winds blew over the soil, breaking it down into fine dust. This was then blown away during severe dust storms.

**Far left** Guipa'go or Lone Wolf, a Native American chief

**Left** Navajo people tend their sheep in Monument Valley, Arizona.

**Right** Bison still live in Yellowstone National Park, Wyoming, because the park is protected from overhunting of its animals.

# WHEAT AND WOOL

Farmers have taken over most of the land in the world's dry grasslands. They have used drier areas for ranching and wetter areas for growing crops. The steppes of Europe and Asia contain Ukraine, one of the 15 countries formed when the former Soviet Union broke up in 1991. This country has rich black soils called chernozem, and Ukraine is often called the "breadbasket of Europe." Ukrainian farmers grow huge amounts of wheat and other crops, including barley, corn, potatoes, and sunflowers. They also lead the world in producing sugar beets.

The pampas, a region much like the prairies of North America,

makes up about a fifth of the South American country of Argentina. Wheat is again a major crop, together with corn and alfalfa, which is used to feed cattle. Crops are grown in the wetter, eastern parts of the pampas. In the drier

**Above** *The productive wheat fields in southern Ukraine*

**Left** *A gaucho herds cattle on the Argentinian pampas.*

**Right** *Farmers use the fertile pastures near Canterbury, New Zealand.*

west, cowboys called gauchos herd cattle on huge ranches. Sheep raising is the chief activity in the cool, semidesert Patagonia region to the south.

The grasslands of southeastern Australia are also important for sheep farming and wheat growing. Many Australian farmers now combine animal raising and crop growing, and most of Australia's wheat now comes from farms that combine the two. Australia has more sheep than any other country, and it leads the world in wool production. Large quantities of wheat and other farm products are now sold to China, Japan, and other Asian countries.

Not far behind Australia is New Zealand, which ranks fourth in the number of sheep and third in wool production. Sheep are reared throughout New Zealand. The country's best known dry grassland region, the Canterbury Plains on South Island, is now the chief grain-growing region.

# Changing Environments

## PAST CLIMATES

In the middle of the Sahara, in southern Algeria, is a plateau (flat-topped upland) called the Tassili n'Ajjer. This area is rich in beautiful rock paintings. Some show animals, such as elephants, giraffes, leopards, and ostriches, which now live only in wet tropical countries. Other paintings show hunters pursuing animals with bows and arrows. Another group of paintings shows nomads herding goats and sheep.

Experts believe that these mysterious paintings by unknown artists tell the story of the Sahara since the last Ice Age ended about 10,000 years ago. They believe that around 8,000 years ago the Sahara was much wetter than it is today, with grassland, trees, rivers, and lakes. The earliest people in the area were hunters and gatherers, who lived much like the Bushmen in southern Africa. Around 5000 B.C., North Africa began to dry up. Farming peoples replaced the hunters. They wandered the grasslands, seeking pasture for their animals.

The Sahara became a true desert only about 3200 B.C. Because the desert could support only a small population, most people had to move to other areas. One place where there was plenty of water was

*Left*  Rock painting in central Algeria

*Right above*  As North Africa dried up to create the desert, oases such as this one near Dahab in Sinai, Egypt, drew people to them.

*Right below*  The Nile River at Luxor, Egypt, is one of the main sources of water in the desert.

the Nile Valley. Here, about 3100 B.C. people founded one of the world's great early civilizations, Ancient Egypt.

World climates have changed many times in the past, and often these changes have helped shape human history. No one knows exactly why climates change. Some scientists believe that the advances and retreats of the ice sheets during the last Ice Age may have been caused by changes in the tilt of the earth's axis and variations in its orbit (path) around the sun. Today, human activities may also be causing climatic changes.

# SOIL EROSION

The setting up of ranches and wheat farms in the dry grasslands has caused problems. In many areas, too many cattle, sheep, and goats grazed the land. They killed off the grasses and other plants, and the land was laid bare. When the rains came, the water was no longer absorbed by the roots of plants under the surface. Instead, much of it flowed over the surface, washing soil away into streams and rivers. This process is called soil erosion. In some areas, the soil was completely stripped away, creating barren areas called badlands.

When droughts occurred, the bare soil was attacked by strong winds. Droughts occur when the normal patterns of air systems over any area change. Scientists do not understand why these changes occur, so they cannot predict how long they will last.

When strong winds blow across bare land, dry grains of soil are pushed across the surface and broken down into fine dust. Winds lift this dust high into the atmosphere and carry it away. Wind erosion occurred on a large scale in the southern part of the Great Plains in the United States in the 1930s. Farmland in parts of five

**Left** This farmhouse in the Dust Bowl of Texas was abandoned in 1938.

**Right above** A dust storm is sometimes called a dust devil.

**Right below** Gulley erosion can be seen in Wildrose Canyon, Death Valley, California.

states—Colorado, Kansas, New Mexico, Oklahoma, and Texas—were affected and the region became known as the "Dust Bowl." Much of the region's once-fertile soil was blown eastward and ended up in the Atlantic Ocean. One great storm in May 1934 carried more than 300 million tons of soil from the Great Plains to the east coast of the United States.

All of the world's dry grasslands have been damaged by overgrazing and poor farming methods. As a result, some former grasslands that were important food-producing regions have become barren desert.

# CONSERVATION AND RECLAMATION

*Left* An irrigation channel, northeast of Antalya, Turkey, brings water to the land.

*Right above* This winter wheat is growing in an area reclaimed from the desert, in Israel.

*Right below* Sand dunes are stabilized to control erosion.

Natural forces, including running water, winds, and moving bodies of ice, wear away the land. This natural process, called erosion, takes place so slowly that its effects are hard to see in one person's lifetime. Soil erosion, caused by human misuse of the land, is a much faster process. Its effects can often be seen within a few years.

To halt the loss of land in dry grasslands and other regions, scientists have worked out many improved farming methods. In the very dry regions, trees are planted to break the force of the winds. Another way of reducing wind erosion is by covering bare land with plants such as clover. This protects the land against the strong winds, and when it is plowed into the soil, it helps make the soil fertile.

Steeply sloping land is not suitable for farming because water running down the slope will remove the soil. Even on gently sloping land, farmers who plow up and down the slope make furrows for rainwater to run quickly downhill. Instead, farmers should plow around the slope. In some areas, where land is in short supply, people do farm steeply sloping land by building steplike terraces. The terraces are flat and are supported by walls or mounds of earth.

In some barren areas, people reclaim barren land and make it

suitable for growing plants. In some countries, farmers plant hardy grasses and trees in sandy areas, which halts the movement of windblown sand. Then they bring water to the land, through pipes or along channels. Eventually, with this irrigation water, they can grow crops. The water for these projects often comes from rocks far below the surface. Some rocks contain a huge amount of water that has been stored there for thousands of years.

# Glossary

**Air pressure**  Air pressure is produced by the weight of air above us in the atmosphere. High air pressure occurs when cold air sinks downward. Low air pressure occurs when warm air rises.

**Anticyclones**  Large regions of high air pressure.

**Atmosphere**  The layer of air around the earth.

**Caravan**  A train of people and pack animals, such as camels, that cross barren areas. Caravans were an important way of trading goods over long distances.

**Chernozem**  A black or dark brown soil found in steppe regions.

**Cloud**  A mass of tiny water droplets or ice crystals in the air. The droplets and ice crystals are so light that they do not fall to the ground.

**Condensation**  The process by which invisible water vapor turns into a visible liquid form (as water) or a solid (as ice).

**Doldrums**  A zone of low air pressure that encircles the world around the equator.

**Drought**  A long period when rainfall is much lower than usual.

**Equator**  A line of latitude running around the world exactly halfway between the North and South Poles.

**Erosion**  The wearing away of the land by natural processes.

**Evaporation**  The process by which a liquid becomes a vapor or gas.

**Fahrenheit**  Scale used to measure temperature. The freezing point of water is 32 degrees Fahrenheit (32 °F), and the boiling point is 212 degrees Fahrenheit (212 °F).

**Hemisphere**  Half a sphere.

**Horse latitudes**  Two zones of high air pressure that encircle the earth around latitudes 30° north and south.

**Irrigation**  The watering of land by

artificial methods, including canals and ditches.

**Lines of Latitude** Lines running around the earth parallel to the equator. Lines of latitude are measured in degrees between the equator (0° latitude) and the poles (90° north and south). Lines of longitude run around the world at right angles to the lines of latitude.

**Middle latitudes** Those parts of the earth between the low latitudes in the hot tropics and the high latitudes around the cold poles.

**Nomad** A person who moves from place to place in search of pasture and food.

**Orbit** The path followed by a heavenly body, such as the earth, around another body, such as the sun.

**Overgrazing** The result of allowing too many farm animals to graze a certain area, so the pastureland is badly damaged or destroyed.

**Pastoral nomad** A person who takes herds of animals from place to place in search of pasture and water.

**Ranching** A system of farming based on raising animals, usually on large farms.

**Semiarid** Partly arid, with scanty rainfall, though not as dry as deserts.

**Stomata** Tiny porelike openings in the leaves of plants through which exchange of gases takes place.

**Temperature** The measurement of how hot or cold something is.

**Trade winds** Two zones of winds that blow from the horse latitudes to the doldrums. The northeast trade winds blow in the Northern Hemisphere. The southeast trade winds blow in the Southern Hemisphere.

**Water vapor** Invisible moisture in the atmosphere.

**Wheat** A member of the grass family that, together with barley, corn, and rice, is a major food.

# Further Information

**Books**

Amsel, Sheri. *Deserts.* Habitats of the World. Austin, TX: Raintree Steck-Vaughn, 1992.

Bernard, Alan. *Kalahari Bushmen.* Threatened Cultures. New York: Thomson Learning, 1994.

Flint, David. *The World's Weather.* Young Geographer. New York: Thomson Learning, 1993.

Hunt, Joni P. *The Desert: Hot & Dry but It's Home to Big Cats, Camels, Coyotes, & More.* Morristown, NJ: Silver Burdett Press, 1994.

Mason, John. *Weather and Climate.* Our World. Morristown, NJ: Silver Burdett, 1991.

Reading, Susan. *Desert Plants.* Plant Life. New York: Facts on File, 1990.

Taylor, Barbara. *Weather and Climate: Geography Facts and Experiments.* Young Discoverers. New York: Kingfisher, 1993.

Twist, Clint. *Deserts.* Ecology Watch. Morristown, NJ: Silver Burdett Press, 1991.

# Index